*C*ontainer homes represent a harmonious blend of sustainability, affordability, and architectural ingenuity. They challenge the conventional norms of housing while addressing some of the most pressing issues of our time, such as environmental impact and affordable living. While there are challenges to overcome, the potential benefits of container houses cannot be overlooked. As society moves towards more sustainable and innovative solutions, container houses are likely to play an increasingly important role in shaping the future of housing.

Container homes have gained popularity as sustainable and cost-effective housing solutions. When it comes to interior decoration, choosing the right materials is crucial to create a comfortable and aesthetically pleasing living space. This book explores various materials that can be used for the interior and exterior decorations of container houses.

Table of Contents

Chapter One

CONTAINER HOUSES

In a world where sustainability and innovation are at the forefront of architectural design, container houses have emerged as a unique and eco-friendly housing solution. These homes, constructed using repurposed shipping containers, have gained significant attention for their versatility, affordability, and minimal environmental impact. This article explores the concept of container houses, their benefits, challenges, and their role in shaping the future of housing.

ORIGINS AND CONCEPT

The concept of converting shipping containers into habitable spaces was born out of necessity and creativity. As global trade increased, surplus shipping containers started piling up in ports around the world. Architects and designers recognized the potential to transform these steel boxes into functional living spaces. The result was the birth of container houses – a marriage of industrial aesthetics and sustainable living.

BENEFITS OF CONTAINER HOUSES

Sustainability: One of the most appealing aspects of container houses is their eco-friendliness. By repurposing existing containers, this housing solution reduces the demand for new building materials and minimizes construction waste. Additionally, using containers decreases the need for traditional construction methods that often generate significant amounts of pollution.

Affordability: Traditional construction can be expensive due to material costs and labor. Container houses offer a cost-effective alternative. The containers themselves are relatively inexpensive, and their modular nature simplifies construction, further reducing labor expenses.

Modularity and Flexibility: Container houses are inherently modular, allowing for easy expansion or downsizing as needs change. Multiple containers can be stacked or arranged in various configurations, offering adaptability to different landscapes and lifestyles.

Durability: Shipping containers are designed to withstand harsh conditions during transit, making them inherently durable. When properly maintained, container houses can withstand extreme weather and last for decades.

Speed of Construction: Traditional home construction can take months or even years. Container houses, on the other hand, can be assembled in a fraction of that time. Since containers arrive pre-fabricated, the majority of the construction work involves modification and interior finishing.

Aesthetic Appeal: Container houses challenge traditional architectural norms and provide a unique and modern aesthetic. Designers have creatively integrated containers into various styles, from industrial to minimalist, resulting in visually striking homes.

CHALLENGES AND CONSIDERATIONS

While container houses offer numerous advantages, they also come with challenges and considerations:

Insulation: Shipping containers are made of steel, which conducts heat and cold. Proper insulation is crucial to ensure a comfortable indoor environment. This can add to construction costs and complexity.

Permits and Regulations: Building codes and regulations vary by location, and obtaining permits for container houses can sometimes be challenging. Zoning laws and homeowner association rules might also impact the feasibility of container homes in certain areas.

Design Limitations: The modular nature of containers can be limiting in terms of interior space and layout. Extensive modifications might be necessary to create larger living areas or incorporate traditional design elements.

Maintenance: While containers are durable, they are susceptible to rust and corrosion over time, especially in humid or coastal environments. Regular maintenance is

essential to prolong their lifespan.

Utilities and Infrastructure: Container houses require proper infrastructure for electricity, plumbing, and waste disposal. Adapting these systems to the constraints of a container can be challenging.

THE FUTURE OF CONTAINER HOUSES

Container houses have already gained popularity in various parts of the world, especially in areas facing housing shortages or seeking sustainable alternatives. As architects and designers continue to innovate, container houses are likely to become more refined and accepted as a mainstream housing option.

In the future, advancements in insulation and building techniques could address some of the challenges associated with container homes. Additionally, the integration of smart technologies could make these houses even more efficient and convenient for residents.

In Essence container houses represent a harmonious blend of sustainability, affordability, and architectural ingenuity. They challenge the conventional norms of housing while addressing some of the most pressing issues of our time, such as environmental impact and affordable living. While there are challenges to overcome, the potential benefits of container houses cannot be overlooked. As society moves towards more sustainable and innovative solutions,

container houses are likely to play an increasingly important role in shaping the future of housing.

Chapter Two

MATERIALS FOR INTERIOR DECORATION OF CONTAINER HOUSES

Container houses have gained popularity as sustainable and cost-effective housing solutions. When it comes to interior decoration, choosing the right materials is crucial to create a comfortable and aesthetically pleasing living space. This article explores various materials that can be used for the interior decoration of container houses.

INSULATION AND FRAMING:

Insulation is paramount in container homes due to their metal structure. Options include spray foam insulation, rigid foam panels, and natural insulators like cork. Proper insulation not only regulates temperature but also reduces energy consumption.

WALLS AND CEILINGS:

Drywall: The most common choice for walls, providing a smooth surface for painting or wallpaper.

Plywood: Offers a rustic look and can be stained or painted. Also, it can be used for both walls and ceilings.

Metal Panels: Embrace the industrial aesthetic by leaving the original container walls exposed or using corrugated mesub-flooring

FLOORING:

Hardwood: Adds warmth and elegance to the space, but might require additional subflooring due to container's structure.

Vinyl or Laminate: Durable, cost-effective, and easy to install, available in various designs.

Concrete: Industrial and modern look, can be polished, stained, or stamped.

WINDOWS AND DOORS:

Sliding Glass Doors: Allow plenty of natural light and create a sense of openness.

Framed Windows: Strategically placed windows can maximize views and ventilation.

Steel Doors: Complement the industrial vibe and provide security.

CABINETRY AND COUNTERTOPS:

Plywood or MDF Cabinets: Lightweight and cost-effective, can be customized with paint or veneer.

Steel Cabinets: Match the industrial style and offer durability.

Quartz or Concrete Countertops: Stylish and durable options for kitchen and bathroom surfaces.

LIGHTING:

Pendant Lights: Add a touch of elegance and serve as statement pieces.

LED Strip Lights: Provide versatile and energy-efficient lighting options.

Skylights: Bring in natural light and create an airy atmosphere.

FINISHES AND TEXTURES:

Brick Veneer: Introduce texture and warmth to the interior.

Wood Paneling: Achieve a cozy cabin-like feel.

Metal Accents: Incorporate elements like exposed steel beams for an industrial look.

Color Palette and Aesthetics:

Choose colors and aesthetics that reflect your personal style. Neutral tones can create a modern, minimalist atmosphere, while bold colors add vibrancy and character.

Furniture and Décor:

Opt for space-saving and multipurpose furniture to maximize the limited space in container homes. Use décor elements like rugs, curtains, and art to personalize the space.

Designing the interior of a container house requires a thoughtful selection of materials that balance functionality, aesthetics, and sustainability. By choosing the right insulation, walls, flooring, windows, and more, you can create a unique and comfortable living space that truly reflects your style and values.

Chapter Three

FACTORS THAT DETERMINE STANDARDS AND QUALITIES OF A GOOD CONTAINER HOUSES

1. Structural Integrity and Material Selection

Structural integrity is crucial for container homes, as they must withstand various loads and environmental factors. When selecting materials, consider the container's original steel structure, and reinforce weak points if necessary. Opt for weather-resistant coatings to prevent corrosion. Consult with an engineer to ensure your design and material choices meet safety standards.

2. Compliance with Building Codes and Regulations

Container homes, or homes built using repurposed shipping containers, are subject to building codes and regulations just like traditional homes. However, compliance can vary depending on your location. It's essential to research and consult local authorities to understand the specific requirements for your area. While container homes offer unique benefits, ensuring their structural integrity, safety, and adherence to zoning laws is crucial for a successful project.

3. Efficient Space Utilization

Efficient space utilization in container house building involves careful planning of layouts, considering multi-purpose furniture, maximizing vertical space, and utilizing storage solutions. Designing with functionality in mind can help make the most of the available area and create a comfortable living space.

4. Functional Design and Layout

When designing a functional layout for a container house, consider your specific needs and lifestyle. Start with defining zones for living, sleeping, cooking, and bathing. Maximize natural light by placing windows strategically. Use multi-purpose furniture to save space and ensure efficient storage solutions. Prioritize insulation for temperature control. It's essential to maintain a balance between aesthetics and practicality to create a comfortable and efficient living space.

5. Plumbing and Electrical Systems

Consider consulting with professionals experienced in container home construction to properly plan, design, and implement these systems. Ensure proper insulation, appropriate fixtures, and adherence to local building codes for safety and efficiency.

6. Environmental Considerations

When setting up a container house, it's important to take environmental considerations into account. Here are a few key points to keep in mind:

Insulation and Energy Efficiency: Containers can become too hot in summer and too cold in winter due to their metal construction. Proper insulation is essential to maintain comfortable temperatures and reduce energy consumption for heating and cooling.

Sustainable Materials: Opt for eco-friendly building materials whenever possible. Consider recycled or reclaimed materials for construction to minimize environmental impact.

Renewable Energy Sources: Integrate solar panels or other renewable energy sources to power your container home, reducing reliance on non-renewable energy and lowering your carbon footprint.

Water Efficiency: Install water-saving fixtures and consider rainwater harvesting systems to reduce water consumption and minimize strain on local water resources.

Natural Ventilation: Design the container layout to allow for natural cross-ventilation, which can reduce the need for mechanical cooling systems.

Waste Management: Implement a waste management plan that includes recycling and proper disposal of construction waste. Additionally, plan for waste reduction and recycling within the container house itself.

Site Selection: Choose a site that minimizes disruption to the local ecosystem and landscape. Avoid sensitive areas like wetlands and forests.

Low-Impact Foundations: Use low-impact foundation systems that minimize excavation and disruption to the land.

Permits and Regulations: Ensure that your container house adheres to local building codes, zoning regulations, and environmental requirements.

Landscaping: Plan landscaping with native plants that require less water and maintenance, helping to blend your container home into the natural environment.

Remember that each location and context might have unique environmental considerations, so tailor your approach accordingly. Consulting with architects, builders, and environmental experts can help you create a sustainable and environmentally conscious container house.

7. Aesthetics and Customization

Aesthetic customization for a container house can involve various aspects such as paint colors, exterior finishes, landscaping, and interior décor. You can choose a theme or style that resonates with you, like modern, rustic, or minimalist, and then select colors, materials, and design elements accordingly. It's a great way to make your container house feel unique and personalized.

8. Quality of Workmanship

The quality of workmanship for a container house depends on factors like the construction crew's skill, attention to detail, and adherence to building codes. To ensure good quality, hire experienced professionals, inspect the work regularly, and communicate your expectations clearly.

9. Longevity and Maintenance

Maintaining a container house is similar to maintaining a traditional house. Regular tasks include cleaning, inspecting for leaks or rust, checking insulation, and maintaining HVAC systems. How often you perform these tasks can depend on factors like climate, usage, and the quality of materials used in construction. Generally, a thorough inspection and maintenance routine every 6 months to a year is a good practice.

Chapter Four

FACTORS THAT DETERMINE THE COST OF CONTAINER HOMES

Container Acquisition: The cost of purchasing and transporting the shipping containers themselves can vary based on their size, condition, and location.

Site Preparation: Clearing, leveling, and preparing the land for construction can add to the overall cost.

Foundation and Installation: Creating a solid foundation for the containers and securing them in place can incur costs.

Insulation and Ventilation: Proper insulation is crucial for temperature control. You might need to insulate the containers and ensure adequate ventilation, which can impact costs.

Structural Modifications: Cutting openings for doors and windows, reinforcing walls, and connecting containers may require professional help and additional expenses.

Utilities: Installing plumbing, electrical systems, and other utilities can contribute significantly to the overall cost.

Interior Finishes: Flooring, wall treatments, fixtures, and furnishings all contribute to the final cost.

Permits and Regulations: Depending on local regulations, permits might be required for various aspects of the construction, which can come with associated costs.

Labor Costs: Labor expenses for design, engineering, construction, and finishing can vary based on location and the complexity of the project.

Customization: The level of customization you desire will impact costs. Basic designs are more affordable, while intricate designs require more effort and resources.

Transportation and Crane Rental: If the containers need to be transported to the site or stacked using a crane, there will be additional expenses.

Maintenance and Longevity: Consider the long-term costs of maintenance and potential modifications or repairs over time.

Land Costs: The price of the land itself is a crucial factor that influences the overall budget.

It's essential to thoroughly research and budget for each of these factors to get an accurate picture of the total cost of building a container house. Consulting with professionals, such as architects, contractors, and builders, can help you make more informed decisions and estimate costs accurately.

THE PHASES OF BUILDING A CONTAINER HOUSE: FROM CONCEPT TO HOME

1. Planning and Design Phase
2. Procurement of Shipping Containers
3. Site Preparation and Foundation
4. Container Modification and Preparation
5. Structural Framework and Assembly
6. Plumbing and Electrical Installations
7. Insulation and Interior Finishing
8. Exterior and Aesthetic Enhancements
9. Testing and Quality Assurance
10. Final Touches and Move-In
11. Post-Construction Considerations

Chapter Five

POST-CONSTRUCTION CONSIDERATIONS FOR CONTAINER HOUSES

Building a container home is an exciting and innovative journey, but the responsibilities don't end once the construction phase is completed. Proper maintenance and ongoing considerations are essential to ensure the longevity, comfort, and functionality of your container house. Let's delve into the post-construction considerations that every container homeowner should keep in mind.

1. Regular Maintenance

Just like traditional homes, container homes require routine maintenance to prevent wear and tear. Regularly inspect the exterior for rust, especially in areas where the containers were modified. Address any rust spots promptly with appropriate treatment and paint to prevent further corrosion. Check for leaks in the roof and walls to maintain a watertight Interior.

2. Insulation Inspection

The insulation of a container home is crucial for maintaining a comfortable indoor environment. Regularly inspect the insulation to ensure it's still in good condition. Address any areas where insulation might have shifted or become damaged. Effective insulation not only regulates temperature but also prevents moisture buildup that could lead to mold and other issues.

3. Plumbing and Electrical Systems

The plumbing and electrical systems in your container home should be regularly inspected to ensure they're functioning properly. Check for leaks, drips, and any signs of water damage in the plumbing system. Test all outlets and switches to ensure they're working correctly in the electrical system. Address any issues promptly to prevent inconvenience and potential hazards.

4. HVAC Maintenance

If your container home is equipped with a heating, ventilation, and air conditioning (HVAC) system, regular maintenance is essential. Change air filters as recommended by the manufacturer, clean ducts, and ensure that the HVAC system is functioning optimally. Proper maintenance will ensure efficient temperature control and good indoor air quality.

5. Exterior Finishes

Over time, the exterior finishes of your container home might require touch-ups or repainting. This not only enhances the visual appeal but also helps protect the structure from the elements. Inspect siding, paint, and any added architectural features for signs of deterioration and address them as needed.

6. Pest Control

Just like any other home, container houses are susceptible to pests. Regularly inspect the exterior and interior for signs of pest infestation, such as rodents, insects, or termites. Implement preventive measures like sealing entry points and using appropriate pest control methods to safeguard your home.

7. Foundation Maintenance

If your container home is built on a foundation, periodic inspection of the foundation is crucial. Look for signs of settling, cracking, or erosion around the foundation. Address any issues promptly to prevent structural problems down the line.

8. Landscaping and Drainage

The landscaping around your container home can affect its longevity. Ensure proper drainage to prevent water accumulation around the foundation, which could lead to moisture issues. Regularly trim vegetation that could potentially damage the structure or obstruct airflow.

9. Energy Efficiency

Continue to focus on energy efficiency even after construction. Consider upgrading to more energy-efficient appliances, LED lighting, and exploring renewable energy sources like solar panels. These measures not only reduce your environmental footprint but also save on utility bills.

10. Adaptability

Your lifestyle and needs may change over time. Consider the adaptability of your container home to accommodate these changes. Keep the layout flexible and plan for potential additions or modifications in the future.

Owning a container home comes with its own set of responsibilities, particularly in the post-construction phase. Regular maintenance, inspection of key systems, and a proactive approach to addressing issues are crucial to ensuring the long-term quality and comfort of your container house. By prioritizing these considerations, you'll continue to enjoy the unique benefits of your container home for years to come.

Container Homes Gallery

Container Homes

Container Homes

Container Homes